# PLANETS

Rebecca Woodbury, Ph.D., M.Ed.

**Gravitas Publications Inc.**

# PLANETS

Illustrations: Janet Moneymaker

Planets
ISBN 978-1-950415-42-7

Published by Gravitas Publications Inc.
Imprint: Real Science-4-Kids
www.gravitaspublications.com
www.realscience4kids.com

**RS4K**

Photo credits: Cover & Title Pg: NASA, Public Domain; Above, NASA, ESA, A. Simon (Goddard Space Flight Center), and M.H. Wong (University of California, Berkeley); P.5. NASA, Public Domain; P.9. Top, NASA, Public Domain; Bottom, NASA, ESA, A. Simon (Goddard Space Flight Center), and M.H. Wong (University of California, Berkeley); PP. 11, 13, 15, 17: NASA, Public Domain

When you go outside, you
might see your neighbors.

Hello, neighbor!

Earth has neighbors too.
Earth and its neighbors
are called **planets.**

We live on
a planet.

# Review: PLANET

To be a **planet,** an object in space must...

**1.** Be big enough to have its own **gravity.** Gravity is the force that holds everything to the surface of a planet.

**2.** Be **spherical** (ball-shaped).

**3.** Move around a sun. **Orbit** is the name for the almost circular path a planet follows around a sun.

Earth is one of 8 planets
that orbit the Sun.

I want to visit another planet!

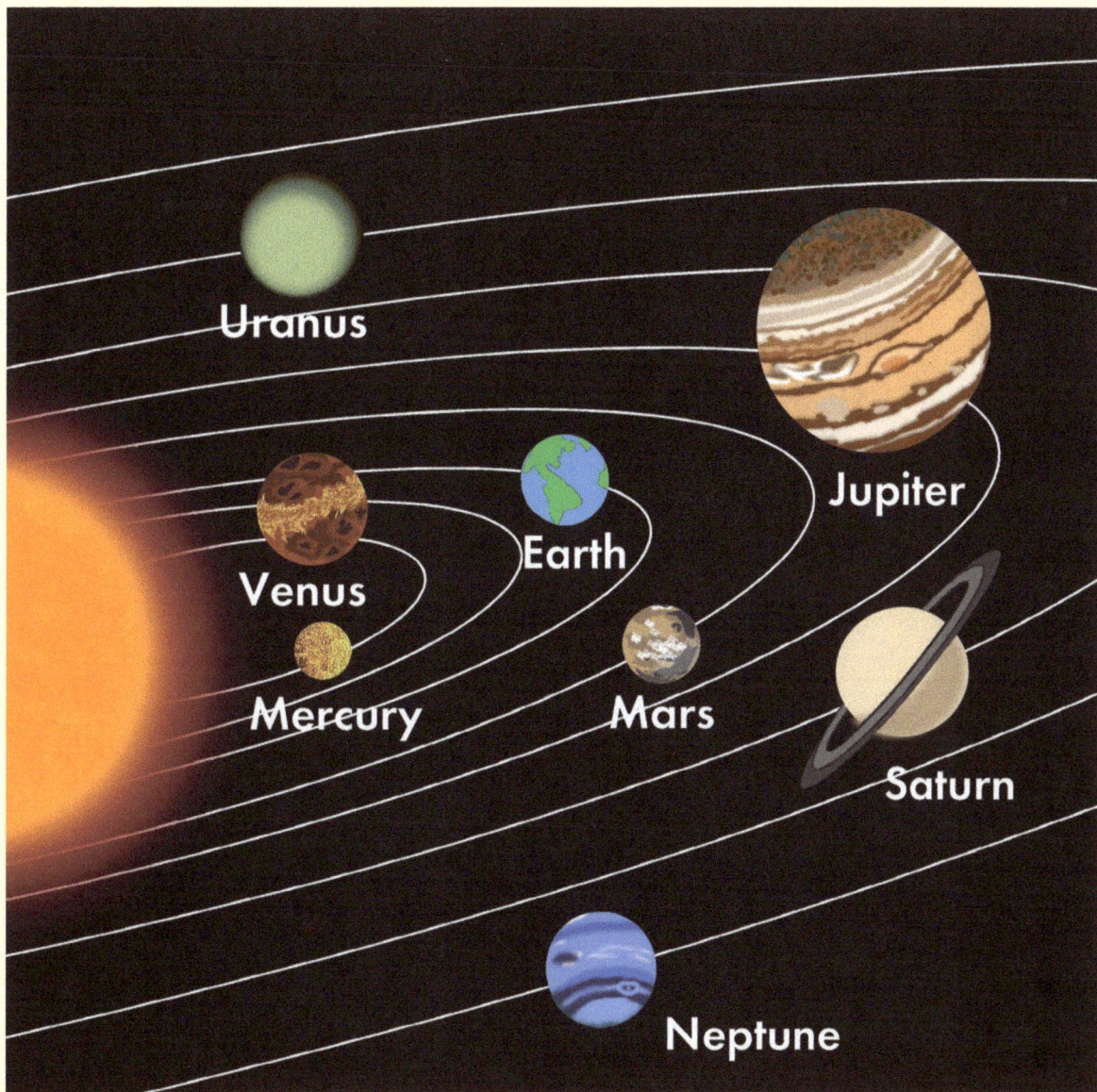

Uranus

Jupiter

Venus

Earth

Mercury

Mars

Saturn

Neptune

There are two different types
of planets that orbit the Sun.
These are **Earth-like planets**
and **Jupiter-like planets.**

Earth

**Earth-like** planets are made of rocks. They are called **terrestrial** planets.

**Jupiter-like** planets are made of gas. They are called **Jovian** planets.

Jupiter

There are four terrestrial planets that are made of rocks. These are **Mercury, Venus, Earth, and Mars.**

Rocks!

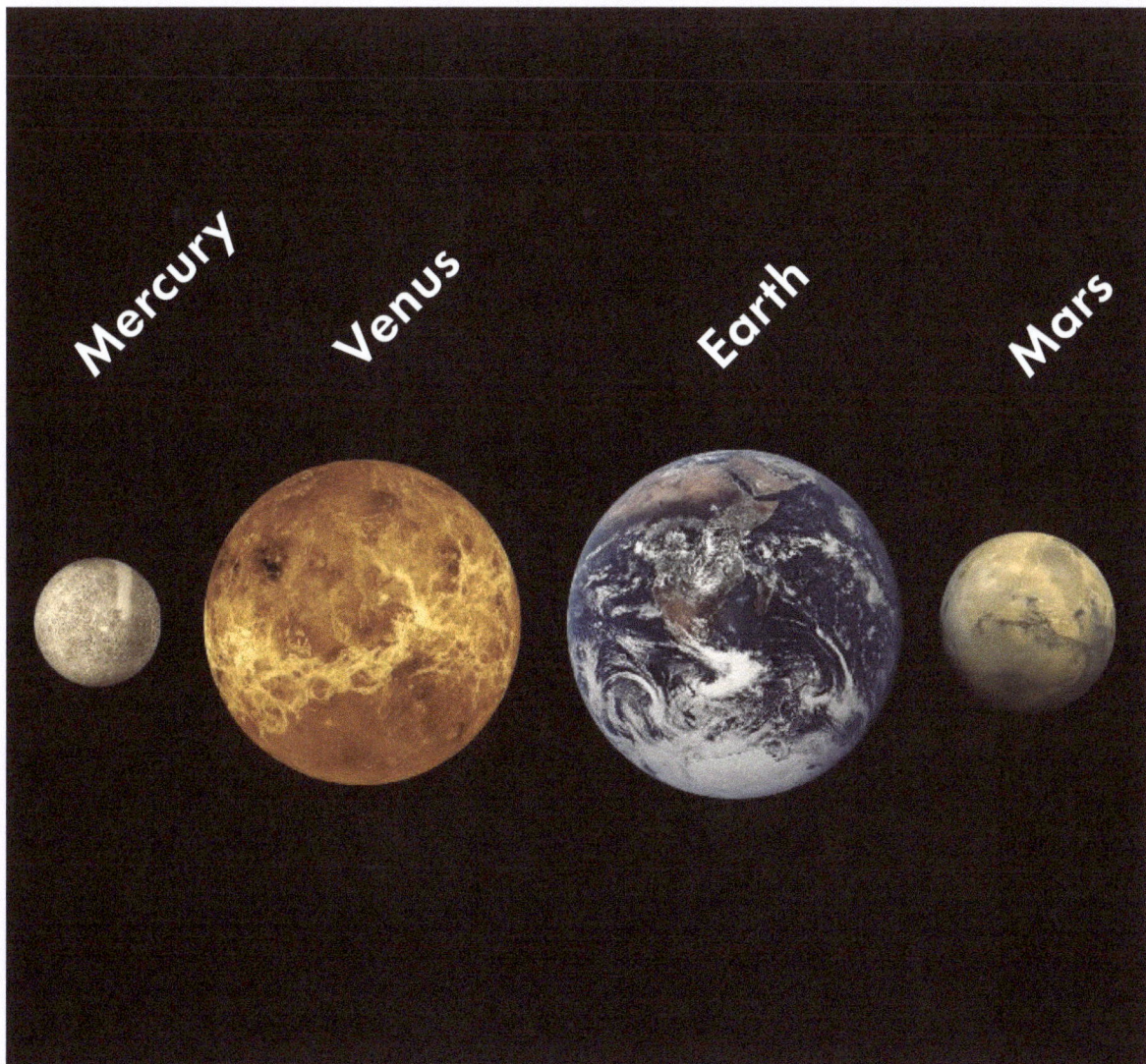

Mercury    Venus    Earth    Mars

There are four Jovian planets that are made of gas. These are **Jupiter, Saturn, Uranus,** and **Neptune.**

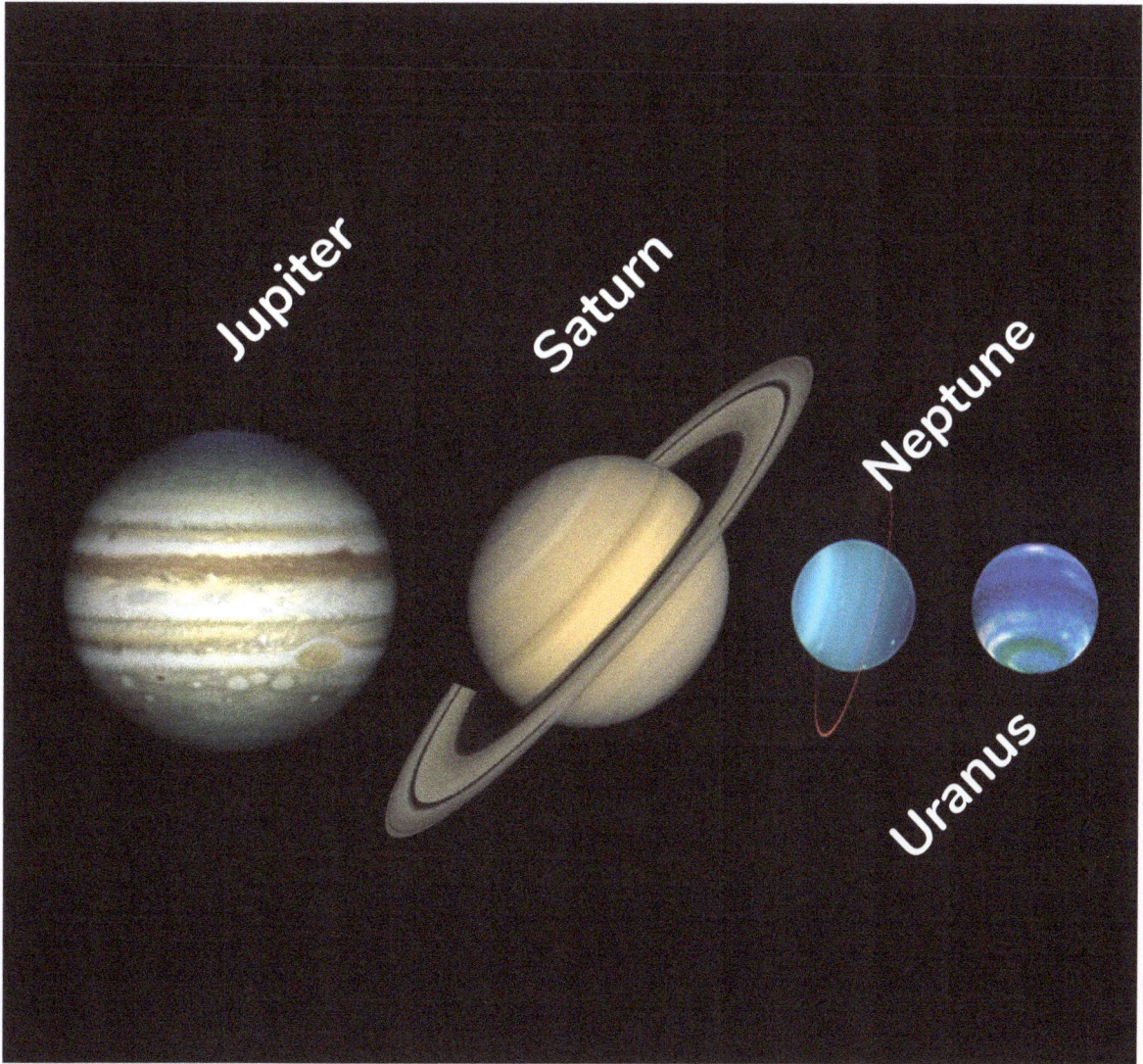

The terrestrial planets are smaller and closer to the Sun than the Jovian planets.

I wonder why?

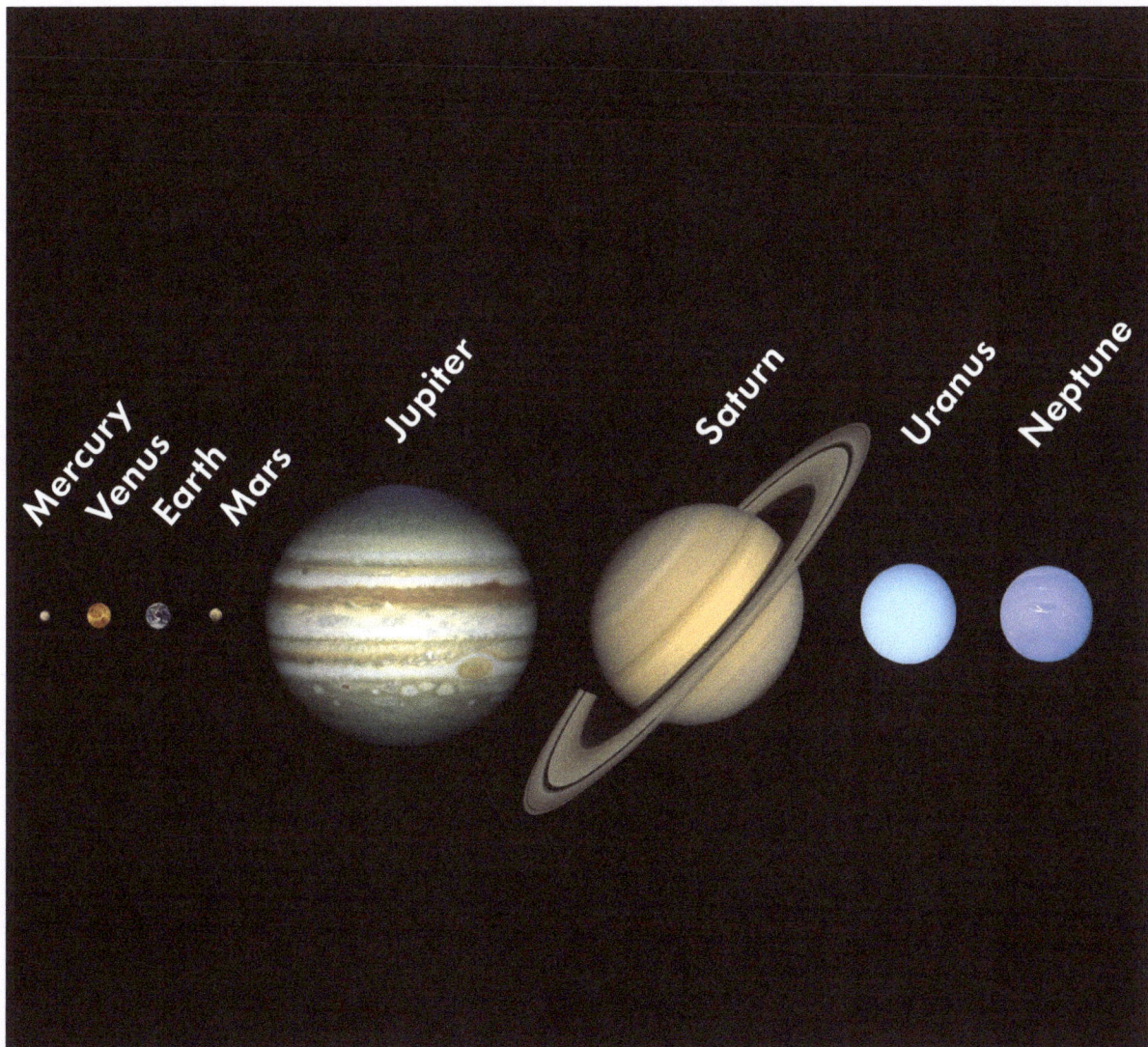

Mercury Venus Earth Mars Jupiter Saturn Uranus Neptune

A **moon** is a natural object that orbits a planet.

- Earth has one moon.

- Mercury and Venus have no moons.

- All the other planets have two or more moons.

I wonder what it would be like if Earth had 2 moons.

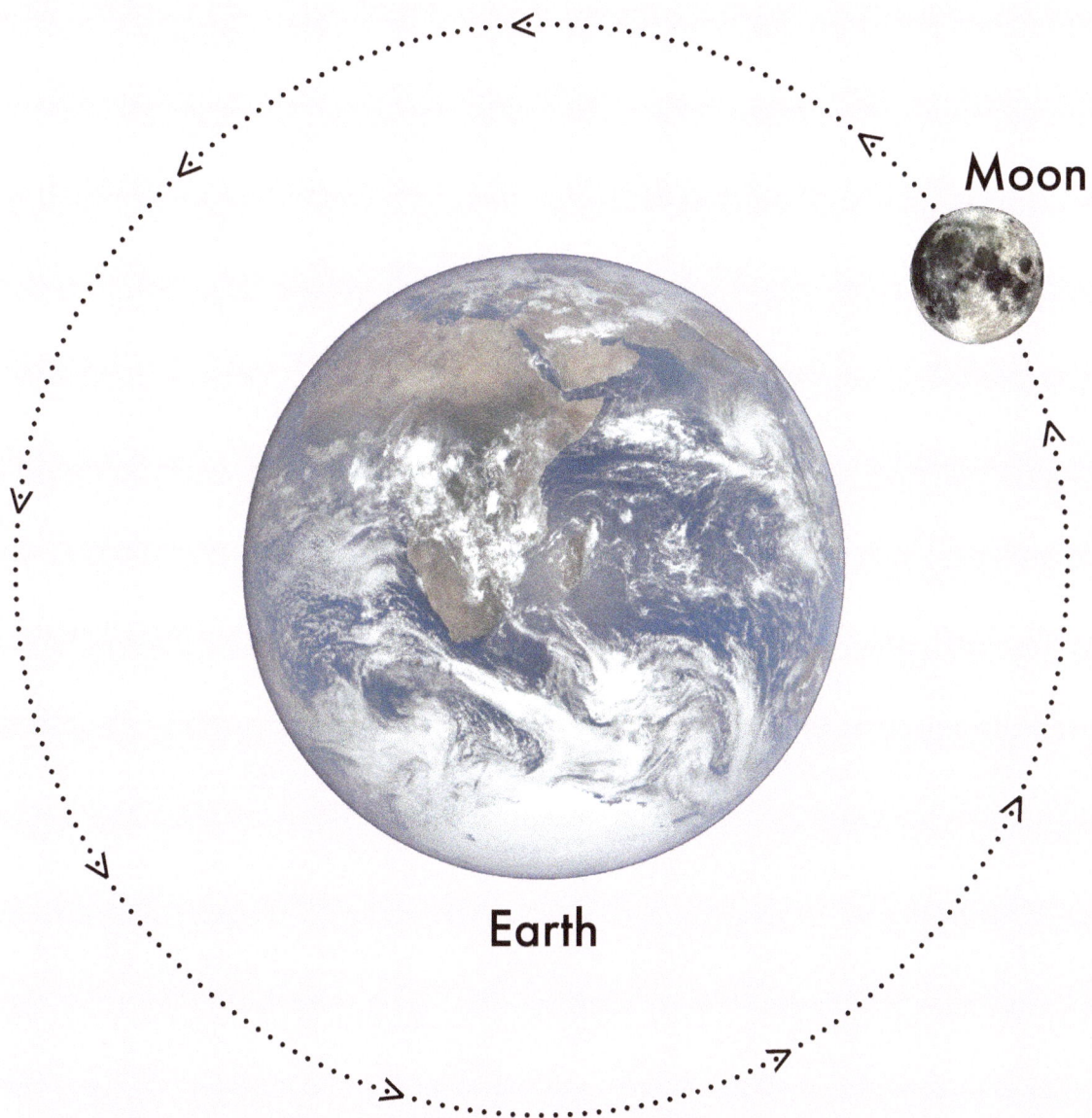

Moon

Earth

## The Moon orbits Earth

Each planet moves in its
own orbit as it circles the Sun.

Good! They cannot crash into each other.

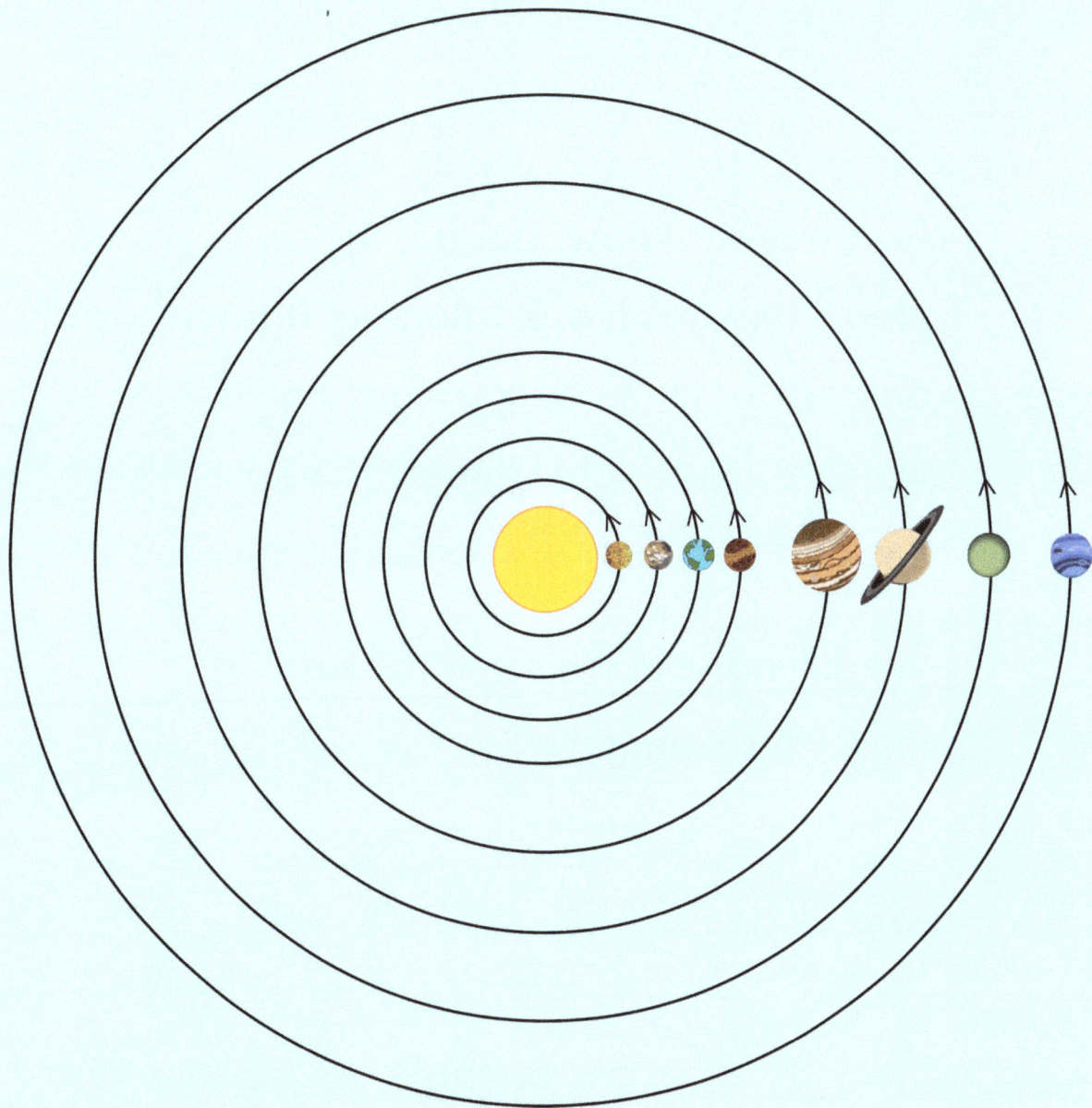

As far as we know, Earth is the only planet that has life. Earth is the right distance from the Sun to have liquid water and the right temperatures. It has air that plants and animals can use.

We still have much more to learn about the planets.

We love Earth!

# How to say science words

gravity    (GRAA-vuh-tee)

Jovian    (JOH-vee-uhn)

liquid    (LIH-kwid)

orbit....(AWR-buht)

planet    (PLAA-nuht)

science    (SIY-uhns)

space    (SPAYSS)

spherical    (SFIR-ih-kuhl)

temperature    (TEM-puhr-chuhr)

terrestrial    (tuh-RES-tree-uhl)

water    (WAW-tuhr)

## Planet Names

Earth    (ERTH)

Jupiter    (JOO-puh-tuhr)

Mars    (MAHRZ)

Mercury    (MUHR-kyuh-ree)

Neptune    (NEP-toon)

Saturn    (SAA-tuhrn)

Uranus    (YOOR-uh-nuhs)

Venus    (VEE-nuhs)